Up Close™

Sea Monsters

PAUL HARRISON

PowerKiDS press
New York

Published in 2008 by The Rosen Publishing Group
29 East 21st Street, New York, NY 10010

Copyright © 2008 Arcturus Publishing Limited

All rights reserved. No part of this publication may be reproduced in any form without permission in writing from the publisher, except by a reviewer.

Author: Paul Harrison
Editor (new edition): Kate Overy
Editor (US edition): Kara Murray
Designer (new edition): Sylvie Rabbe

Picture credits: Bridgeman Art Library: page 5, top, page 6; Chris Harvey-Clark: page 13, bottom; Corbis: page 7, top; Natural History Museum: page 9, top and bottom; Nature Picture Library: page 11, bottom, page 13, bottom, page 16, top, page 17, page 18, top; NHPA: page 8, page 16, bottom, page 19, top, page 21, bottom; Science Photo Library: front cover, page 5, bottom, page 10, page 13, top, page 15, top, page 21, top, page 22; Topfoto: page 4, page 10, page 12, page 15, bottom.

Library of Congress Cataloging-in-Publication Data

Harrison, Paul, 1969–
 Sea monsters / Paul Harrison.
    p. cm. — (Up close)
 Includes index.
 ISBN 978-1-4042-4224-1 (library binding)
 1. Marine animals—Juvenile literature. 2. Marine animals, Fossil—Juvenile literature. 3. Sea monsters—Juvenile literature. I. Title.
 QL122.2.H36673 2008
 591.77—dc22

                                                    2007033482

Manufactured in China

CONTENTS

- 4 WHAT IS A SEA MONSTER?
- 8 ANCIENT MONSTERS
- 10 SEA SERPENTS
- 12 SUCKERED IN
- 14 SWALLOWED WHOLE
- 17 DEEP-SEA MONSTERS
- 20 WHAT WAS THAT?
- 22 GLOSSARY
- 23 FURTHER READING
- 24 INDEX

WHAT IS A SEA

Legends of sea monsters have been told for as long as people have spent their evenings swapping stories. Seafarers have always come back with terrifying tales of mysterious creatures. There must be something about being at the mercy of the waves that messes with the mind—or is there something more to these unlikely myths?

The biggest sea monsters are blue whales. They can measure a scary 100 feet (30 m) and weigh a massive 187 tons (170,000 kg).

STRANGE WORLD

Not all the blame for these tall tales lies with sailors, though. People who regularly took to the seas were used to seeing creatures such as whales or sharks—but they tended not to be the people who could write. The passengers were generally the literate ones, as they could afford education and travel, but most of them only went to sea once. Imagine seeing a whale for the first time and not knowing what it was— you would probably think that you had seen a monster, too! Other scholars who wrote of sea monsters may not have been to sea themselves at all, but were merely recounting tales that had been told to them.

MONSTER?

TO BOLDLY GO

What we have to remember is that sailing could be a terrifying experience. Battling stormy seas in leaky wooden boats was a constant reminder of how close a neighbor death could be. Explorers pushed the boundaries of the known world farther and farther, and with this came a natural fear of the unknown—and being lost and exhausted is bound to play havoc with your nerves!

MONSTER MAPS

Many old sea maps were covered with pictures of monsters thought to be lurking in the depths. Before embarking on a voyage into the fearsome seas, sailors would consult the maps and plot their course to avoid bumping into a giant octopus or sea serpent. Monsters also made the maps look interesting and exotic.

The carcasses of strange sea creatures that wash up on the shore are known as *globsters*. Sometimes even the experts don't know what they are!

SCARE STORIES

There was a time when stories of sea monsters were used for more than just entertainment. Some early traders deliberately spread rumors of sea monsters as a way of scaring off the competition from following their highly profitable trade routes. Why bother going all the way to a distant port when there was a good chance you'd end up as a light lunch for a creature from the deep?

DEM BONES

In the past, superstition was rife. Not everything has a simple explanation, so the supernatural replaced most things which today can be accounted for by science. Dinosaur bones on land helped fuel the belief in dragons—so why shouldn't monsters live in the sea, too? Such beliefs were bolstered when rotting carcasses of whales or other large marine life were washed onto the shore. The decayed corpses looked like nothing anyone had ever seen and were consequently rumored to be monsters.

LURKING BENEATH

As a rule, octopuses will eat almost anything they can get their hands on. Their long tentacles are very strong and their suckers are touch-sensitive, releasing chemicals which they use to stun or poison their prey. Imagine seeing one of these hiding beneath the waves!

NEW DISCOVERIES

Over two-thirds of the Earth's surface is covered by water, and new species of sea dwellers are being discovered every day. In 2003, an Arctic fisherman came across an enormous squid, almost 20 feet (6 m) in length, with huge protruding eyes and spiked tentacles. Scientists believe that such a species could grow to up to 40 feet (12 m)! Perhaps there are more monsters out there than we think ...

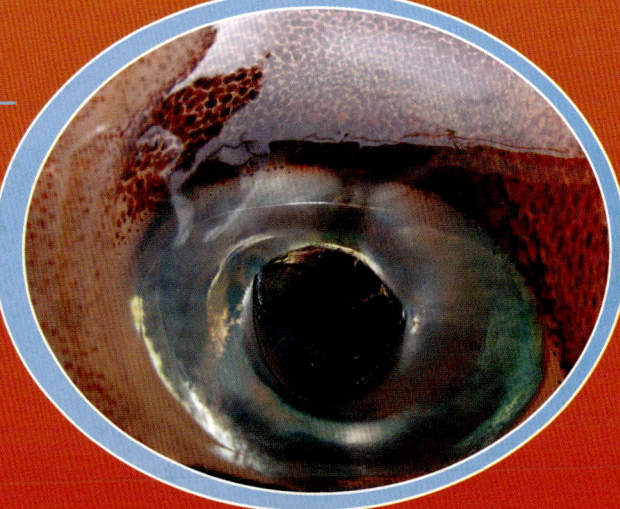

ANCIENT MONSTERS

Life developed in the sea long before it did on land, so it comes as no surprise that there have been some pretty scary creatures floating around our oceans in the past. The top predators on the planet used to live in the water—here are some of the most impressive.

Sharks first appeared on Earth around 400 million years ago.

SNAP, SNAP!

One type of ancient sea monster that is still with us is the crocodile. The largest crocodile today—the estuarine, or saltwater, crocodile—grows to over 20 feet (6 m) in length. This is impressive enough, but it's positively tiny compared to its ancient relatives. Some early forms of crocodile were absolutely huge—up to 39 feet (12 m) long!

TOOTHLESS TERROR

Over 360 million years ago, one fish not to be messed with was Dunkleosteus (dun-klee-OS-tee-us). At around 18 feet (5.5 m) long, this armor-plated monster was bigger than today's great white shark. Although Dunkleosteus didn't have teeth—each jaw had a hard bony edge instead—it was still the top predator of its day, and was more than happy to eat anything that crossed its path.

MESOZOIC MAULER

One of the most fearsome creatures ever to roam the oceans was the Liopleurodon (ly-oh-PLOOR-uh-don). It terrorized the seas during the Mesozoic era, around 150 million years ago. At around 49 feet (15 m) in length, and with a ferocious reputation to match, it was more than big enough to be a worry to everything else in the ocean—even unwary dinosaurs paddling along the shore.

NESSIE

Of course, one of the most famous monsters in the world is Nessie, rumored to inhabit Loch Ness in Scotland. Like many other lake monsters around the world, it is thought that Nessie is a living plesiosaur (PLEH-see-uh-sor)—a marine reptile from the time of the dinosaurs. Nessie spotters often claim to have seen a long neck and a humped back, which is just what a plesiosaur looked like. Could this mean some ancient reptiles still exist?

SEA SERPENTS

The classic sea monster is, of course, the sea serpent. There have been reports of gigantic snakelike terrors in the oceans for hundreds of years, but what's the real story behind this particular menace?

STATESIDE SERPENT

Chesapeake Bay, off the coast of Virginia and Maryland, is reputedly home to a large sea serpent nicknamed Chessie. In 1982, a man named Robert Frew spotted Chessie and managed to capture it—on videotape, that is. The tape went off to the Smithsonian Institute to be examined but, unfortunately, the quality was too poor for scientists to say exactly what it was that Frew saw. So is Chessie real, or is it really something more mundane—a seal or a manatee perhaps?

INLAND SERPENTS

You may not even have to go to sea to spot monstrous serpents—some people believe that certain lakes are home to their own monsters. The most famous lake monsters are Nessie from Loch Ness, Ogopogo from Lake Okanagan in British Columbia, Canada, and Champ from Lake Champlain between Vermont and New York. It has been claimed that these lakes may have underground passages to the ocean, allowing the monsters to meet other sea monsters and breed.

In 1840, the crew of the ship *Pekin* were convinced they'd spotted a sea monster. It turned out to be a big lump of seaweed. Could this be another explanation for sightings of monsters?

FABULOUS FISH

One thing is beyond doubt—one kind of sea serpent does exist. Don't believe it? Well then check out exhibit A, the oarfish. At around 29 feet (9 m) long, and with a red dorsal fin running the length of its body, the oarfish looks quite unlike any other species of fish. Could this unusual beast be what sailors have been spotting for all these years?

SNAKES ALIVE

Of course, there are actually snakes that swim in the sea, too. Some of these sea snakes have flattened bodies, which help them swim, but also make them look odd. Could reports of the river-dwelling anaconda of South America—which grows to over 29 feet (9 m) in length—have convinced some people that such monsters live in the ocean, too?

SUCKERED IN

Take a peek at an octopus or a squid. All those suckery legs writhing around like a nest of snakes is enough to give anyone nightmares. Imagine how you would feel seeing a gigantic version of one of those bearing down on you! No wonder sailors got nervous ...

There's plenty of evidence of really big squid—sperm whales are often found with huge sucker marks on their skin from fights with these massive creatures.

KRAKEN

The most notorious of the ancient, squidlike monsters was the kraken, and there have been many tales of this giant creature rearing out of the sea and dragging boats down beneath the waves with it. It was claimed, in the 1700s—by the bishop of Bergen no less, so you'd hope he was telling the truth—that the kraken was as big as an island. Now that would make a scary vacation destination!

ANCIENT TALES

Seafarers have long told tales of giant octopus and squid attacking their ships. It's such an old theme that these episodes have made it into literature, too—most famously in Jules Verne's *Twenty Thousand Leagues Under the Sea*. Of course, there's little reliable documentary evidence that such attacks ever occurred, but that doesn't mean such monsters don't exist.

GIANT OCTOPUS

Lusca is the name given to a type of giant octopus that is supposed to live in the Caribbean. Legend has it that these monsters are around 131 feet (40 m) long—around 10 times bigger than the documented size for giant octopuses. There's little proof that luscas exist, of course, but a series of old photographs showing a mysterious body washed up in the Bahamas might suggest they are out there ...

REAL-LIFE MONSTER

Today, the closest thing to the kraken that we know of is the giant squid. Some scientists estimate that it could grow up to around 43 feet (13 m) long —that's about the length of a bus. There is some evidence, however, of a colossal squid that could possibly be twice the size of the giant squid— that's a lot of calamari!

SWALLOWED

The fear sailors had of sea monsters was two-fold. Firstly, the monster might sink the ship, leaving the sailors either to drown or be cast adrift to a slow death. Secondly, it might have a taste for a sailor snack and eat them. But how likely is it that a person could get eaten at sea?

GENTLE GIANTS

There are many different types of whale, from humpback whales, which grow to over 40 feet (12 m), to killer whales, which are already 8 feet long at birth. The biggest whales are difficult to spot, as they tend to live in the deepest parts of the ocean. On the rare occasions that whales have eaten people, it tends to be because the person harpooned them first! Whales are placid by nature.

WHOLE

The throats and stomachs of most whales and sharks are too small to swallow a human.

BIG FISH

Given the size of a basking shark's mouth, it's easy to see why sailors might steer clear. Basking sharks are the second biggest fish in the ocean; on average they filter enough water to fill an entire swimming pool in just one hour! But they are not aggressive creatures, and spend most of their time near the surface of the water—hence their nickname, Sunfish.

THE LEGEND OF JONAH

The idea of being swallowed by a sea monster is ingrained in myths, legends, and even religion. The Bible tells a story of a man called Jonah who was swallowed by a large fish. According to the legend, he lived inside the fish's belly for three days and three nights, before being regurgitated—and lived to tell the tale!

TINY FOOD

Whales certainly look big enough to eat a person—but generally, they prefer to dine elsewhere. Many whales eat tiny creatures called krill, which they trap in comblike plates of bone in their mouth, called baleen. Other whales do have teeth, and the biggest of these is the sperm whale. Interestingly, there is a tale that a sailor called James Bartley was swallowed alive by a sperm whale during the 1890s and survived for 16 hours in the whale's stomach!

DEADLY TASTER

Of course, sailors may well have seen people eaten by some types of shark. Great white sharks and tiger sharks do eat people—but it's rare. Usually sharks only attack people when they mistake them for some other kind of food!

DEEP-SEA MONSTERS

As we have seen, not all sea monsters are figments of people's imaginations. In truth, the oceans are home to some of the strangest animals you could imagine—and some of the oddest are found at the very bottom of the sea.

FANG-TASTIC

Some of the scariest teeth to be found in the ocean belong to the viperfish—they're so big they can't even fit inside the fish's mouth. Fortunately, the fish is only 10 inches (25 cm) long, so it's not a real threat to anything on the large side.

In the deepest parts of the ocean, the water pressure is equivalent to having 50 jumbo jets piled on top of you!

17

LURING THEM IN

The bottom of the ocean is so far from the surface that sunlight can't get down all that way. Instead, sea creatures there produce their own light. The aptly named anglerfish uses a glowing ball of light hanging from its dorsal fin to attract small fish, like an angler with a baited hook. When the fish get too close, the anglerfish suddenly springs to life and gulps down the hapless victim.

STRANGE FISH

Because of the pressure, only very specific fish can survive in waters this deep. Soft-bodied animals such as jellyfish and sea anemones thrive, while surface-dwelling fish such as dolphins stay away. It works both ways though—most deep-sea creatures would perish if brought to the surface.

BIG MOUTH

The gulper eel has two of the biggest advantages for a deep-sea predator—a huge mouth and an expanding stomach. The gulper eel's already large mouth can actually unhinge, or dislocate, allowing it to swallow prey as big as itself. Fortunately, its stomach can stretch just as much to accommodate the eel's sizeable lunch.

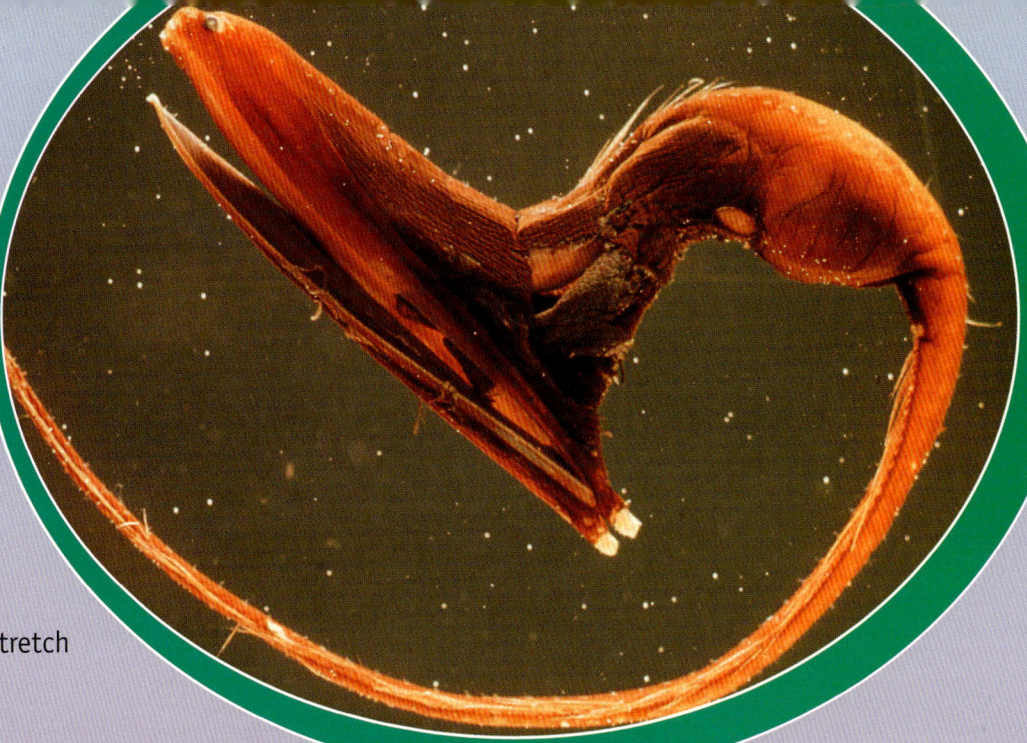

Some scientists believe that up to 90 percent of deep-sea fish can produce their own light.

JEEPERS CREEPERS

The Greenland shark's shining eyes make it look more menacing than it really is—actually it's a harmless creature. This rare shark is one of the few that live in Arctic waters, where it can be found in both deep and shallow water, depending on the time of year. It is also one of the largest sharks alive. The light in its eyes is not produced by the fish itself, but by a parasite which lives on the shark's eyes.

WHAT WAS THAT?

The only way to tell for sure whether there are sea monsters out there is to find them. Unfortunately, that's easier said than done. There's a lot of water out there—it covers two thirds of the planet—and these sea creatures can be very elusive.

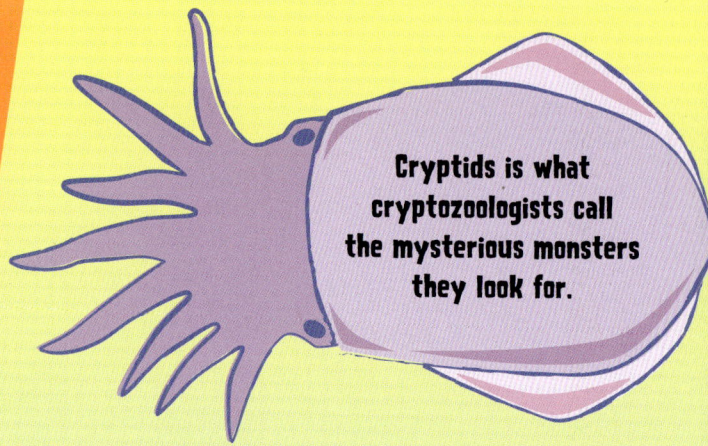

Cryptids is what cryptozoologists call the mysterious monsters they look for.

MEGA SURPRISE

It seems very likely that there are big animals out there that we don't know about. And finding them seems to be a matter of luck. Until 1976, no one knew that the megamouth shark existed, and then one got caught accidentally by a research vessel. If something as big as this 16-foot (5 m) shark can stay hidden for all this time, perhaps, just perhaps, there really are some big monsters out there waiting to be found.

BEACHCOMBING

Sometimes scientists don't have to go looking for interesting creatures. Instead, the sea washes them up, or fishermen net them by mistake. Generally, these creatures have been dead for a long time and are badly decomposed. In 1977, a Japanese trawler hauled what looked like a dead plesiosaur from the waves. There was great excitement until scientists pointed out that this is exactly what basking sharks look like as they rot.

SCIENTISTS OR NUTS?

The practice of looking for animals that most people consider to be either myths or extinct is called cryptozoology. Many people scorn cryptozoologists, declaring them not to be real scientists. However, some cryptozoologists do use very strict scientific criteria in their work.

HIDE AND SEEK

Most of the sea remains unexplored, mainly because it's very difficult to study such a large area. Also, the sea bed isn't nice and flat; instead it's made up of mountains and trenches—some more than 7 miles (11 km) deep—so there are plenty of places for creatures to hide.

GLOSSARY

CARCASS (KAHR-kus)
The dead body of an animal or bird.

COLOSSAL (kuh-LO-sul)
Huge, enormous, or gigantic.

CRYPTOZOOLOGY (krip-tuh-zoh-O-luh-jee)
The study of creatures that are generally considered to be non-existent or extinct.

DUNKLEOSTEUS (dun-klee-OS-tee-us)
An enormous prehistoric fish, now extinct.

ELUSIVE (ee-LOO-siv)
Difficult to catch.

KRILL (KRIL)
Small shrimplike sea creatures that many bigger fish eat.

LIOPLEURODON (ly-oh-PLOOR-uh-don)
Huge prehistoric fish from around 150 million years ago.

MANATEE (MA-nuh-tee)
A whalelike mammal.

MESOZOIC (meh-zuh-ZOH-ik)
A prehistoric period that began about 250 million years ago.

NOTORIOUS (no-TOR-ee-us)
To be known for bad things.

PARASITE (PER-uh-syt)
An organism that lives on an animal or plant but doesn't help it survive.

PLESIOSAUR (PLEH-see-uh-sor)
A marine reptile from the time of the dinosaurs.

SUPERSTITION (soo-pur-STIH-shun)
An irrational belief that makes you behave in a certain way.

TRAWLER (TRAH-ler)
A boat that catches fish with a large net.

22

FURTHER READING

CHASED BY SEA MONSTERS: PREHISTORIC PREDATORS OF THE DEEP
by Nigel Marven and Jasper James
New York: DK, 2004

THE BOOK OF SEA MONSTERS
by Nigel Suckling and Bob Eggleton
New York: Overlook Press, 1998

ENCYCLOPEDIA PREHISTORICA: SHARKS AND OTHER SEA MONSTERS
by Robert Sabuda and Matthew Reinhart
Cambridge, MA: Candlewick Press, 2006

SEA MONSTERS
by Stephen Cumbaa
Tonawanda, NY: Kids Can Press, 2007

SEA MONSTERS
by Aaron Sautter
Mankato, MN: Coughlan Publishing, 2006

WEB SITES

Due to the changing nature of Internet links, PowerKids Press has developed an online list of Web sites related to the subject of this book. This site is updated regularly. Please use this link to access the list:
www.powerkidslinks.com/upcl/seamons/

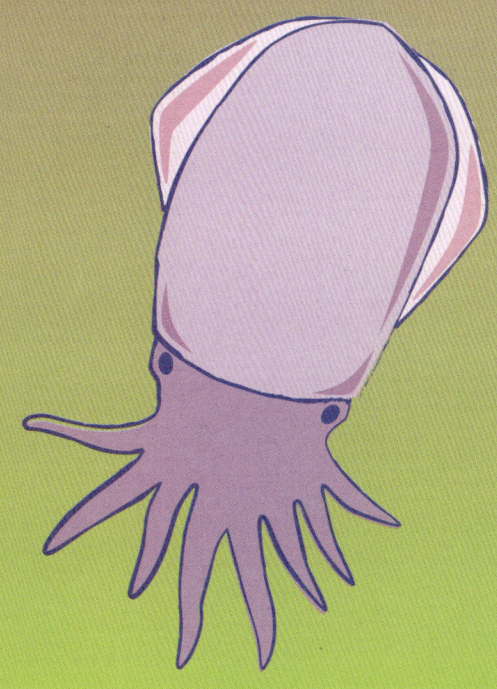

INDEX

A
anaconda 11
anemones 18
angler fish 18
attacks 13, 16

B
baleen 16
basking shark 15, 20
blue whale 4
bones 7

C
crocodiles 8

D
dinosaurs 7, 9
dorsal fin 11
dragons 7

E
eels 19
explorers 5

F
fishermen 20

G
giant octopus 13
giant squid 13
great white shark 9, 16
Greenland shark 19
gulper eel 19

J
Jonah 15

K
kraken 12, 13

L
legends 4, 15
Loch Ness 9, 10

M
maps 5
megamouth shark 20
myths 4, 21

N
Nessie 9, 10

O
oarfish 11
octopus 12, 13

P
plesiosaur 9
predators 8, 9

R
reptiles 9

S
sailors 4, 5, 11, 14, 16
saltwater crocodile 8
scholars 4

scientists 10, 20, 21
serpents 10, 11
sharks 4, 9, 15, 16, 19, 20
snakes 10, 11
sperm whale 12, 16
squid 12, 13
stories 4, 6

T
teeth 16, 17
tiger shark 16
travel 4
trawler 20

V
viperfish 17

W
whales 4, 7, 15, 16

24